# Little Inventors In Space!

Inventing out of this world

by Katherine Mengardon
and Dominic Wilcox

# Credits

Words **Katherine Mengardon**
Drawings **Dominic Wilcox**
Design **Kathryn Corlett**

Publisher **Michelle I'Anson**
Editors **Sarah Woods & Karen Midgley**

With thanks to Suzie Imber, Planetary Scientist, the Canadian Space Agency, the Natural Sciences and Engineering Research Council of Canada (NSERC), Wyn Griffiths, Willem Postema and James Fletcher for their contribution.

This book was made possible thanks to the hard work of our Little Inventors team: Emilie Harrak, Will Evans, Jill Bennison, Ellie Birkhead, Phoebe Martin, Suzy O'Hara, Gareth Lloyd and Chelsea Vivash.

Thank you to our Little Inventors all across the world, whose imaginations inspire us every day, and to all the Magnificent Makers and partners who help us bring them to life!

With a special mention to Nick, Natacha and Rudy Coates, Justine Boussard, Holly and Maya Mataric, Leila Harrak, Max and Maggie Evans, Seren Lloyd, Ruby and Sally O'Hara, Dillon Corlett and Nikol Misiura.

**Little Inventors®** is a registered trademark of Little Inventors Worldwide Ltd.

Images © Little Inventors
Text © Little Inventors
Drawings © Dominic Wilcox

*"Never be limited by other people's limited imaginations."*

**Mae Jemison**
Astronaut

## Published by Collins

An imprint of HarperCollins Publishers
Westerhill Road, Bishopbriggs,
Glasgow G64 2QT
www.harpercollins.co.uk
© HarperCollins Publishers 2020

**Collins®** is a registered trademark of HarperCollins Publishers Ltd.

A catalogue record for this book is available from the British Library.

Printed by GPS Group, Slovenia.

ISBN is 978-0-00-838290-2

10 9 8 7 6 5 4 3 2 1

**MIX**
Paper from
responsible sources
**FSC™ C007454**

FSC
www.fsc.org

## Chapter 1

Get your inventing brain in gear...

# Inventing out of this world

# What is Little Inventors?

We are a team that believes children have really amazing ideas and that they're worth listening to.

We have received tens of thousands of invention ideas from **children just like you** all across the world.

William, age 11, sent us his invention... 'Space homes'

We look at every single idea that comes through and each time it confirms that we are right – the way you look at the world is **unique!**

So we ask **professional artists, designers and makers** to bring your most ingenious ideas to life, through objects, animations or 3D images.

Photo by Spencer Barclay

It was made into an amazingly detailed model by miniature artist Spencer Barclay from Little Canada.

And then our mission is to **share them online or in real life exhibitions** and shout about how great they are! We want to remind everyone that thinking like a child can truly change the world.

Space homes is on display in a museum and you can also see videos of it on littleinventors.org!

# Inventing starts with an idea...

Ideas can happen anywhere, any time, you just need to follow that thought!

Made real by artist **Ella Merriman**.

Photo by Katherine Mengardon

You might be **outside**, daydreaming and thinking about what game to play...

Hali, age 8, invented the Pogo pencil, to draw in playgrounds as you jump!

...or you might be thinking about how to **get somewhere**...

Alexander, age 8, invented the Triple decker bus.

Made real by **Chris Folwell**, AKA Dabble Dabble.

Photo by Sharjah Children Biennial

...or perhaps you want to make a **boring task** more interesting...

Maja, age 8, invented the **sweet walker** to encourage her sister to keep going when going hiking!

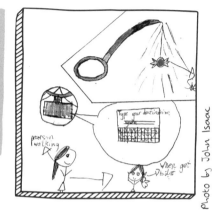

Photo by John Isaac

Made real by artist **Alex McKenzie.**

...or maybe there is something you want **help** with?

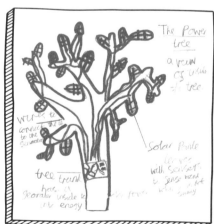

Photo by Richard Kenworthy Photography

Honey, age 10, came up with the Power tree. A pretty way to make energy from the sun!

Made real by **Peter Simon Coyle** from Octo Design.

There really are **no limits** to where your imagination can take you.

# Inventing can take you places!

Connor, age 11, sent his idea for our Inventions for space challenge...

His **Space Boots Imprint** invention allows astronauts to leave their own personalised footprints on the Moon!

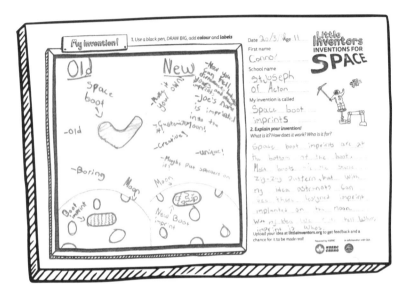

His invention was chosen to be made real by footwear engineer **Chris Bellamy**!

Photos by Chris Bellamy

Photo by Chris Bellamy NSERC

And the **Canadian Space Agency** thought it was such a great idea that it was one of two top winners!

Connor was invited to take part in a **downlink** from the International Space Station and got to chat with astronaut **David Saint-Jacques!**

And he even got to see his invention orbiting Earth!!

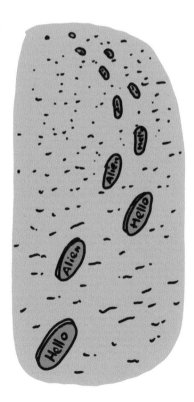

## Chapter 2

Mysterious, infinite, so many possibilities...

# Space is the place

# Space to dream!

Over the centuries, we have learnt a lot about our own planet. But space **still remains mostly unknown** and full of mysteries: about what is out there, and how it was created – the perfect canvas for new ideas and dreams!

It's also **a very difficult place for us humans to live in**. Outside the cocoon of the Earth's atmosphere, temperatures can be incredibly hot or terribly cold; there is no air to breathe and a lot of radiation that could make us very sick!

Maybe it is exactly this mix of **hopes and challenges** that make space and inventing such good companions.

# Our journey into space

We might have dreamt about space for thousands of years, but **it's less than sixty years** since going into space became a reality. And there have been a lot of firsts during those years!

**1957** – **Sputnik 1** is the very first man-made satellite to take off!

**1961** – **Yuri Gagarin** is the first human in space

**1963** – **Valentina Tereshkova** is the first woman in space!

**1969** – First **Moon landing!**

**1976** – First images of **Mars!**

**1990** – First giant **space telescope**

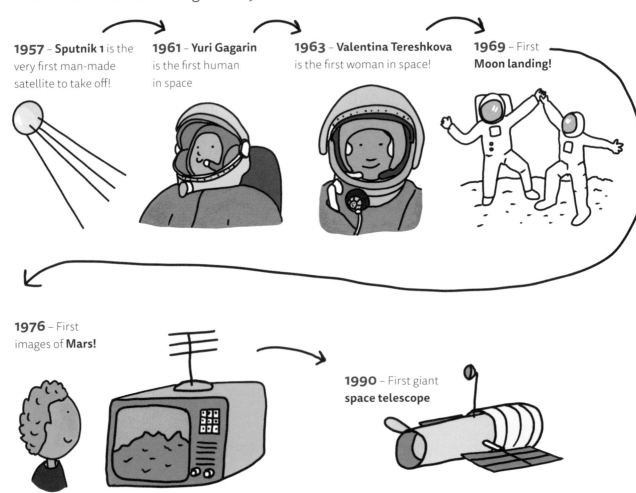

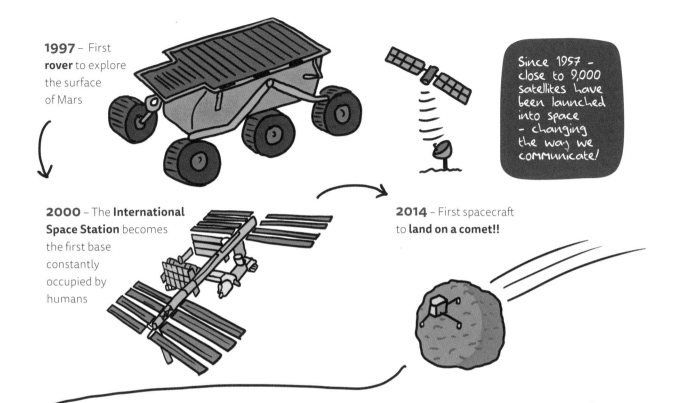

**1997** – First **rover** to explore the surface of Mars

Since 1957 – close to 9,000 satellites have been launched into space – changing the way we communicate!

**2000** – The **International Space Station** becomes the first base constantly occupied by humans

**2014** – First spacecraft to **land on a comet!!**

**2018** – The first Little Inventors **Inventions for space challenge** in Canada!!

More than 80,000 children participated, 30 inventions were made, and you can see some of our favourites in this book!

And what next?
A human base on the Moon or even Mars, before taking on the rest of the universe!

# Inspired by space

There are many inventions that started life helping astronauts in space, but that we now use (almost) every day!

## Freeze-dried food

Astronauts need food that is nutritious, lightweight and keeps well!

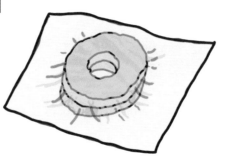

Freeze drying keeps 80% of the nutrition value at only 20% of the normal weight!

## Trainers

The rubber soles of trainers are moulded just like space helmets, for extra comfort.

## Camera phones

Scientists worked on creating small but powerful cameras for use on spaceships – they are now used in a lot of smartphones!

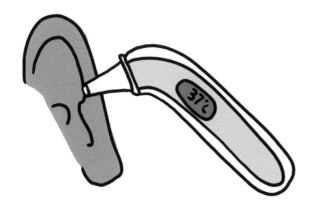

## Ear thermometers

They use infrared technology to take your temperature instantly. They're fast and safe and were first developed to monitor the health of astronauts in space!

## Memory foam mattresses

You know those mattresses and pillows that keep your shape and are extra comfy? They use material developed to make the journey of space pilots more bearable as they sit in the same position for hours.

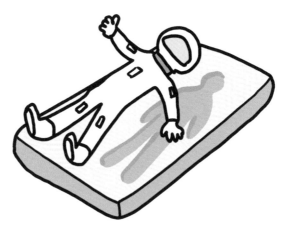

## Artificial limbs

Memory foam technology is also used to create limbs that look more life-like. But that's not all! Robotics developed to control space vehicles are also used to improve the way the limbs move!

# Ready for take-off?

This book will help you explore space with your inventing goggles on, from looking at the sky to preparing to meet life on other planets. Think about the challenges of life in space so you can come up with your very own space inventions...

Fly away

In orbit

The Moon!

Mars!

...with our **Chief Inventor Dominic Wilcox** as your guide!

Could you imagine Dominic's stained glass driverless car on the Moon?

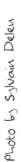
Photo by Sylvain Deleu

# Dominic's Top Tips to start inventing

## Seriously... not!

Space can feel like a very serious subject, but inventing doesn't have to be! If you **put aside rules or laws of physics**, who knows what ideas you might come up with?

Having ideas is essential, figuring out how to make them work can come later!

## Let your mind play!

Inventing is another way to say you are giving your mind **licence to play**. It can wander, explore, experiment with ideas and see how they fit together. And the more play the better!

## Get inspired...

Every single idea doesn't have to be completely new.

Many inventions are actually **improvements** on existing inventions!

So be curious and find out about other ideas, who knows how this will inspire your own?

## Doodle away

You don't always need to know what you're drawing... In fact, **you might like to doodle** and then imagine what you can see in your doodle!

Check out inventions by other Little Inventors like you at littleinventors.org.

# What would you like to do in space?

Draw yourself. Don't forget your astronaut suit and helmet!

I'd like to go to...

I would go with...

The first thing I want to discover is...

# Your name in the stars!

Join the stars whose names begin with each letter of your name –
this is your very own constellation!

Capella · Betelgeuse

· Arcturus

· Pollux

Xamidimura · Zosma

Fornacis · Lesath Terebellum

Etamin · Unukalhai Yildun · Grumium · Ogma

Diadem · Kitalpha · Sirius · Vega

· Izar · Muscida

Jabbah · Wesen · Hadar

· Naos · Rastaban

Quetzalcoatl

What shape can you see in your star name?

## Chapter 3

Destination out of this world...

# The sky is definitely not the limit

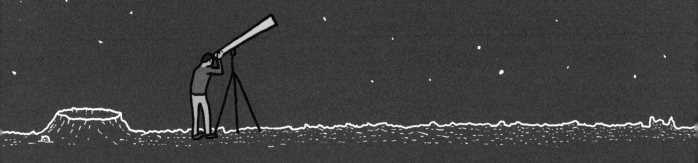

# Look up, look up!

As we go about our daily lives, we understand what it is we see around us. Cars, houses, our school... we know what they are, how big they are, and how far away from us they are.

But then, we look up at the night sky...

What we see is on a **completely different scale**. There is literally no end to what is above our heads. Staring at the sky and space beyond is **the closest thing to seeing infinity**.

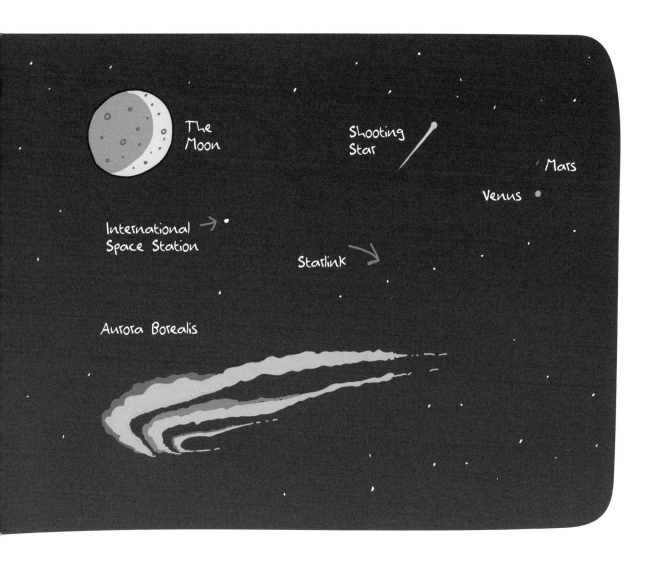

An infinity full of comets, stars, planets, galaxies... All the things that make up this **wider natural world**. It also includes most of the objects that we humans have sent to space during the last 60 years.

# Now you see me...

What you might see in the night sky just by looking up (or with the help of a telescope) are literally millions of stars, as well as the Sun and Moon, but that's not all...

There are five planets you can spot with your naked eye: **Mercury, Venus, Mars, Jupiter** and **Saturn** – and while they might appear as small dots in the sky, their true sizes tell another story!

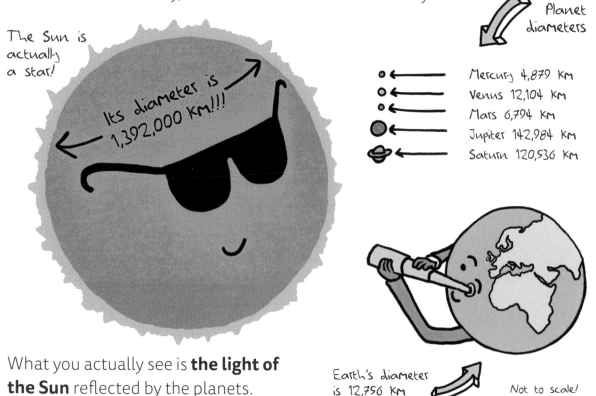

Planet diameters

The Sun is actually a star!

Its diameter is 1,392,000 km!!!

Mercury 4,879 km
Venus 12,104 km
Mars 6,794 km
Jupiter 142,984 km
Saturn 120,536 km

What you actually see is **the light of the Sun** reflected by the planets.

Earth's diameter is 12,756 km

Not to scale!

# What stars are made of...

...in short, very hot gas!

Sirius, Betelgeuse and Rigel are some of the brightest stars you can see!

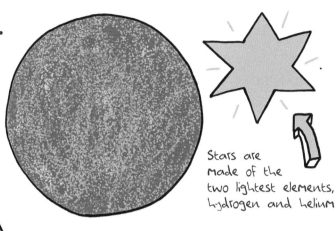

Stars are made of the two lightest elements, hydrogen and helium

# A long time ago in a galaxy far, far away...

These stars, as well as over 150 billion other stars, are part of our galaxy. But you might also be able to spot the Andromeda galaxy, our largest neighbour!

# Light pollution

All the artificial lighting we use is causing a **glow** that can make it difficult to observe the night sky. There are now designated dark sky spots across the world to try and reduce light pollution.

# Man-made!

There are other things you can see in the night sky that are much closer to us. These are things we have sent there.

The International Space Station (ISS) circles the Earth **sixteen times each day**, but you can see it best at dusk or dawn!

# ...Now you don't!

**Space is immense** and what we can see is only a fraction of what is there above our heads.

Some planets, galaxies and stars are too far away for us to see, but there are also other things that don't capture light and are invisible to us...

## Gravity

This is the **force that pulls things towards** all the physical objects (like planets and stars) in the universe. It's what stops us from floating off into space when we are standing on the Earth.

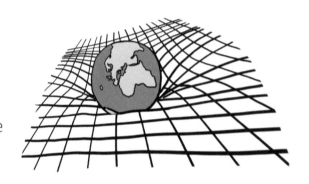

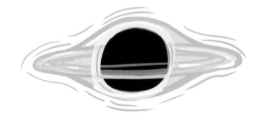

## Black holes

These suck up everything around them, **even light!**

## Dark matter

This is the stuff that **fills the space between stars and planets**, kind of holding the universe together. It doesn't reflect light at all, so we can't see it.

# Space junk

It's not only the Earth that humans have polluted... Over the years, we have been sending more and more engines, devices and satellites into space, but what happens when they break? NASA estimates that **close to 500,000 pieces of space rubbish (or debris)** have been left floating around the Earth's orbit since we started going into space.

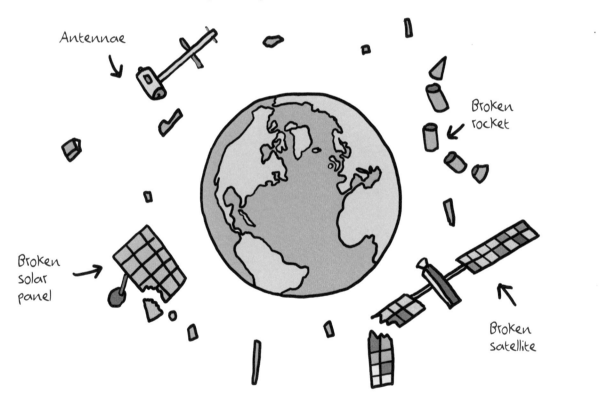

Antennae

Broken rocket

Broken solar panel

Broken satellite

In fact, scientists think that **this cluttering** is one of the biggest problems that needs to be tackled in space right now. And their ideas are pretty out there...

# Clear some space!

Old satellites, rocket shards, bits of comets... all floating around, creating a very real crashing risk for satellites or space engines.

One of the ways of **disposing of space junk** is to bring it back from orbit. Because of the speed it travels, it burns as it re-enters the atmosphere!

**The Pac-Man system** will grab a large bit of rocket and take it out of orbit!

100 kg of old rocket junk!

Burn it!

What system could you imagine to bring junk out of orbit?

Write your Ideas

Another idea is to create a rocket that collects **old satellite parts** and adds them to itself to recycle them into a new satellite. How neat!

How could you recycle space junk?

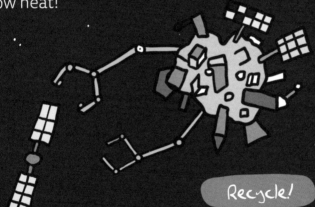

Recycle!

The **TAMU Space Sweeper** with Sling-Sat would use the energy gained from capturing one piece of debris to 'hop' to the next one, to avoid using a lot of fuel.

How would you power your debris machine?

Power up!

# Blast off with your first invention!

Think about what you'd like to invent first. It could be something to stop light pollution or help get rid of space junk, whatever inspires you!

chief Inventor

Ideas are like little bursts of light in our brain, so catch your moment in the Sun and get inventing!

My night sky inspiration

Name it

How it works

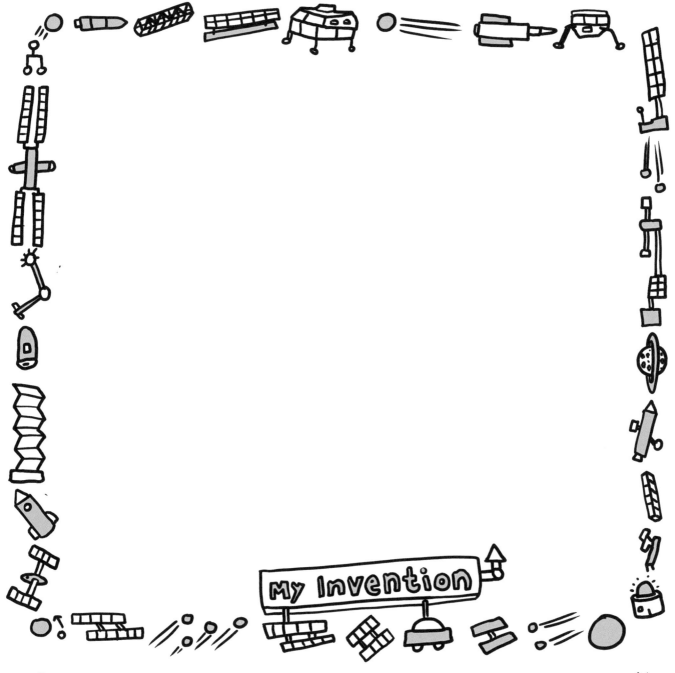

my invention

Draw BIG, use colours and add labels!

Now share it on littleinventors.org!

# Here's looking at YOU!

You have started to conquer the night sky!

SPACE JUNK SPOTTER

Keep shining a light on all the known and unknown parts of space!

You might want to continue to learn about the different planets, stars and other man-made creations in our galaxy and beyond.

But it's always worth thinking about **the impact that we humans have on our environment**, on the Earth and around it.

We need to treat our wider universe as our very own home!

# From the minds of our Little Inventors...

# The repair drone

**Jack, age 12**
Kingston,
Canada

*"It goes outside the International Space Station and repairs stuff so astronauts don't have to!"*

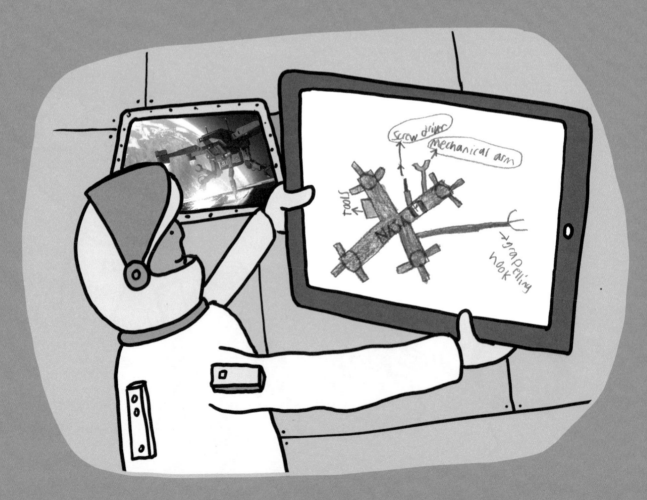

...to the skills of our Magnificent Makers

The repair drone was brought to life as a digital image, created by **Cristian Vasquez at Atomhawk.**

"To me, this idea really stood out and I could really envision it happening. These days, drones are commonly used and I really felt inspired to push the technology in this alternative reality."

# Shooting star chaser!

Rocks that move so fast as they enter the atmosphere that they heat up and glow are known as meteors or shooting stars.

**Meteoroids** are small rocks that fly through space. Some of them will cross the Earth's path. Most of them burn out before they reach the ground.

The ones that do land on Earth are called **meteorites**. These are really valuable as they are pieces of actual planets or stars that can teach us more about the universe.

**Some people actually hunt these meteorites** – generally in deserts or the Antarctic where the small black rocks are easier to spot!

## Chapter 4

Telescopes, satellites and time machines...

# Reach for the stars

# The future is out there...

When it comes to space, this mostly unknown world has captured our limitless imaginations for a very long time.

For thousands of years, travellers on land and sea have used the positions of stars to navigate. Greek and Roman astrologers began reading the stars to explain our personalities and predict the future, and they named the planets after their gods and goddesses.

We have also looked to the skies for **answers to the bigger questions**, and seen them as home to unseen powers.

And what of the future of space? **Science fiction stories** tell us what life in space could be like.

There is truly no end to how space inspires us!

# Blast from the past

**Astronomy** is the study of stars, planets and space. Stars and planets were written about **as far back as 3,500 years ago** and references to them were found all across the ancient world, from Viking territories to Egypt and further east in China, Australia and even Polynesia.

Humans noticed the movement of the Sun, Moon and stars and understood that they were linked to day and night, and also to the seasons.

Astronomy is also a story of invention, starting with a **triple magnifying glass** invented in the 1600s by Dutch glassmakers. This was improved on by **Galileo**, who decided to turn it towards the Moon, and it became the first telescope.

Throughout the centuries, astronomers have continued to observe the skies with more and more sophisticated devices.

But of course it hasn't stopped there!

Learning about space and how it works has led us to many more discoveries and inventions.

**1609 – Galileo's telescope** enlarged objects up to twenty times!

Magnifico!

**1668 – Isaac Newton** improved the telescope by adding reflecting mirrors, reducing its size but not its power

Dazzling!

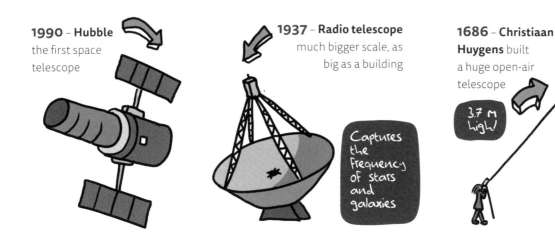

**1990 – Hubble** the first space telescope

**1937 – Radio telescope** much bigger scale, as big as a building

Captures the frequency of stars and galaxies

**1686 – Christiaan Huygens** built a huge open-air telescope

3.7 m high!

# Swooping satellites

The more we have understood about space, the more we have looked to **use it to make our lives better**.

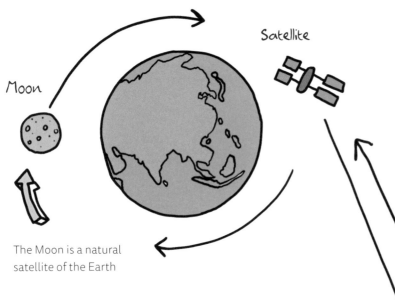

Moon

Satellite

The Moon is a natural satellite of the Earth

Scientists realised that **if we placed satellites in orbit**, they would circle the Earth constantly.

Because satellites are high above us, they can **'see' large parts of the Earth** at any one time, in a way that we can't on the ground. They also have a better view of space than we do!

Every day when we **use GPS** on a mobile phone to check a map, we actually travel to space and back thanks to communication satellites!

space Traveller

Satellites can help predict **weather conditions, and the impact of climate change**. They can measure the amount of ozone and carbon dioxide in the atmosphere and how much energy is absorbed and released on Earth.

They also monitor **volcanoes** – so we know when they might erupt!

Other satellites face space, checking for dangerous solar rays, or exploring asteroids and comets to help us learn more about the history of stars and planets.

They may look for **evidence of water on Mars** or capture close-up pictures of Saturn's rings!

The International Space Station is a really a **science lab in the sky!**

Testing Medicines and Materials

Growing human tissue cells

# Sci-fi or tomorrow?

And who knows what else we will invent when we learn more about space? **Lots of books and films** explore what our future might be when we spend more time in other worlds...

Imagine who you might meet on your space exploration. What invention could help you understand them, with or without words?

A universal translator!

Write your Ideas

who?

What would be a useful tool to have in space? What power would it have?

A screwdriver that fixes anything!

Sometimes you may want to explore safely - can you think of ways to stop others noticing you?

Nothing to see here!

What could you do if you needed help?

Don't panic!

Where would you go if you could travel in time?

1800 1900 1000 1700 1600 1200 1100 1500 1300 1400

A time machine!

# Sirius-ly cool inventing!

So what will YOUR invention be? Something to explore planets from, a new satellite to solve problems, or something to make the impossible possible?

My future of space inspiration

When you want to invent something, you can start by observing what's around you, or letting your imagination fly!

Name it

How it works

MY INVENTION

Draw BIG, use colours and add labels!

Now share it on littleinventors.org!

# Mission star explorer complete!

Observe, imagine and dream – who knows what the future will bring!

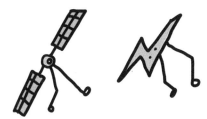  

When space captures your imagination, see where it takes you!

It might be that you want to know all the facts about **gravity, orbit, space technology or the different atmospheres** on each planet...

...or you might be more interested in **dreaming up what life in space might be like** and come up with your own vision.

Both are great ways to explore space. Who knows, they might help with the next big space discovery!

# From the minds of our Little Inventors...

# The sound collector

> Could you hear a pin drop on Mars? That's what we would like to know!

**Eben, age 6**
Calgary, Canada

*"You can see the piece labelled sound collector. It collects sound and has great grip on its robotic legs. It helps so the astronauts don't need to get anything else."*

sound collecter

air vent

on batton off

button

robotic Legs

good grip

# ...to the skills of our Magnificent Makers

Eben's invention captured the imagination of our **Chief Maker Gareth Lloyd**! The two had a video chat to talk about how it would look and how big the robot should be – about the size of a small dog!

*"I used parts from a fire engine toy and a Meccano kit for the body, but I didn't just want this model to look like it worked I wanted it to ACTUALLY work!*

*I thought it would be cool if the boom could point towards motion so that it could then record sound. This would be useful when looking for aliens.*

*It is quite uncanny when you walk up to it as it turns to hear you!!"*

sound collecter

air vent

robotic legs

good grip

Photo by Gareth Lloyd

# Future imaginer!

Some people make a living out of inventing the impossible!

**Science fiction writers** mostly tell stories about a future where imagined scientific and technological advances transform human (or non-human!) society, often featuring space and life on other planets.

The strange thing is that reading books that were written a long time ago can sometimes feel like they were predicting our present! Although not always...

What future can you imagine for us in space?

## Chapter 5

Bacteria, blobs and other aliens...

# Is there anybody out there?

# Crazy little thing called CHNOPS!

How does life begin? This question has kept humans wondering and arguing for centuries. But whatever people believe, there is a simple fact that links every living thing together...

They are **all mostly made of the same elements!**

Earth happens to have the right combination of all of these, but that's not all. Its distance from the Sun, its composition, the presence of an atmosphere... **all of these factors are key to life existing on planet Earth.**

But what if these conditions could also be found on other planets or in other galaxies, in one form or another?

We have wondered **what other life forms are out there** – would they look like us, or be completely different? Would we even recognise them as living things?

One thing is certain, they would need to follow the rule of CHNOPS!

# Needle in the space stack

So for life to sprout on another planet, it would need plenty of CHNOPS! But what does that really mean?

First and foremost a planet needs a **star** (like our Sun) that is neither too big nor too small, just at the right distance to provide the planet with enough **energy** for **water** to exist in liquid form – the source of all life! If it was further away, the water would be ice, too close and it would soon evaporate.

It also needs **nutrients** for life to feed off, in the form of minerals for example, and an **atmosphere** to provide air to breathe and to protect the planet from dangerous space radiation.

Looking for life beyond Earth has kept space scientists busy for years.

So far, they haven't found much evidence, but there are an estimated **40 billion planets** (known as exoplanets) in our galaxy alone which could possess the right conditions for life...

The challenge is to actually find them!

Engineers and designers have to find ways to spot tiny planets in the **intense brightness of nearby giant stars...**

The Kepler telescope spent 9 years searching space for exoplanets

One idea is **Starshade**.

The shape of the petals on this giant 'shade' could help **bend the light** for a better chance of catching a glimpse of these potential planets.

# Life's what you make it...

Life can take incredibly varied forms on Earth – from tiny microbes to giant trees, and all the plants and animals in between!

But each planet will have its own special composition and that **means life could take very different forms.**

Life on Earth is mostly made of **carbon**, but **silicon** is a very similar element that could form living things.

# Hello, who are you?

Imagine you meet a real life alien!
What sort of questions would you ask them?

What's your name?

How old
are you?

What language do
you speak?

How do you move?

Do you have
a family?

Any pets?

Draw your alien

What planet do you come from?

Is it big, small, cold, hot?

Are there oceans?

Draw the planet they come from

What is your home like?

What do you do every day?

What do you do for FUN?

Draw their home!

# Living it up!

Whatever life forms are out there, what an incredible experience it might be to meet them. So what could you invent to make that easier, safer, or simply more fantastic?

Our differences are what make us interesting!

My alien inspiration

Name it

How it works

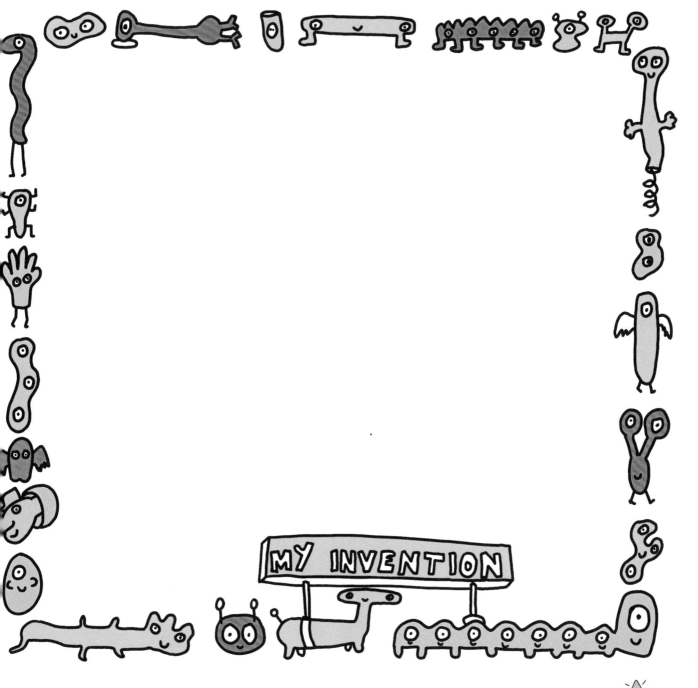

MY INVENTION

Draw BIG, use colours and add labels!

# Live and let live!

**What a blast it could be to meet our neighbours!**

With its **creature-friendly atmosphere**, oceans and plentiful resources, our blue planet is a pretty special place in the universe.

But ultimately it is very hard to believe that we are alone, so it's less of a matter of **if** we'll ever find life out there, but more a matter of **when**. Who knows, it might even be within your own lifetime!

And once this happens, who knows what we will learn and how it will change the way we live?

After all, we will also be aliens to those living creatures so maybe, just maybe, we should try and be on our **best behaviour** so we can impress them!

# From the minds of our Little Inventors...

# Space-imals

**Ella, age 11**
Nepean, Canada

*"This is a space suit for pets, cats mainly!"*

Every day we live with creatures who are very different from us – our pets! There are an estimated 88 million pet cats on Earth, but not a single one in space. What if we could take them along? Who knows, maybe aliens are more cat than human!

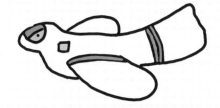

## ...to the skills of our Magnificent Makers

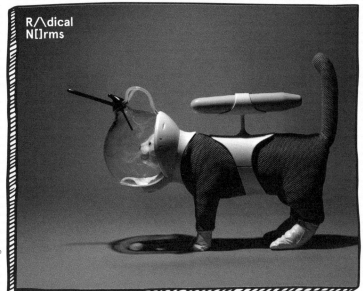

R/\dical N[]rms

Photos by Radical Norms

Ella's cat space suit was brought to life by makers **Koby and Daniel from Radical Norms**, a design research studio in Ottawa.

*"We started by making a 1:1 cardboard cat model so we could plan our suit, and created part templates so we could 3D print them. The rest of the suit is a snug-fitting soft shell for comfort!*

*Zooki inspected the suit and wasn't so keen on the tail part..."*

...a little help from Zooki the cat!

# Space protector!

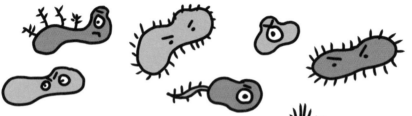

You might think that having to deal with **possible alien life invasions** is something that only happens at the cinema... well not quite.

Part of the job description is to protect the Earth from alien life should it exist... though it's mostly about making sure that spaceships are **safe and squeaky clean**, to avoid the risk of sending our Earth germs to other planets, or bringing any unwanted new space germs back here!

## Chapter 6

Rockets, spacewalks and some light years...

# Come fly with me

# Up, up and away!

Space travel is something that we hear so much about, yet it is not exactly like taking a train, a car or a quick cycle ride around the block.

In reality, it is something that is still very new! For example, **only twelve people have made it to the Moon**, and no one has been there since 1972!

The first rocket was sent into the stratosphere about sixty years ago, and since then, our human visits have mostly reached **only a few hundred kilometres above our heads**, just outside of our atmosphere.

Some spacecraft may have travelled much further than us, but humans are still literally light years away from reaching **deep space**.

Space is simply **on an enormous scale** and doesn't play by our rules...

# A small step for space-kind

Would you believe that the International Space Station is as far from Earth as Sunderland is from London?

About 400 km!

Chief Inventor

That's where I'm from!

Which is quite a distance for us to travel, but **absolutely minute** in space distance!

Space is so huge, that scientists have had to come up with a new truly **astronomical unit (or au)** of measure based on the distance between Earth and the Sun.

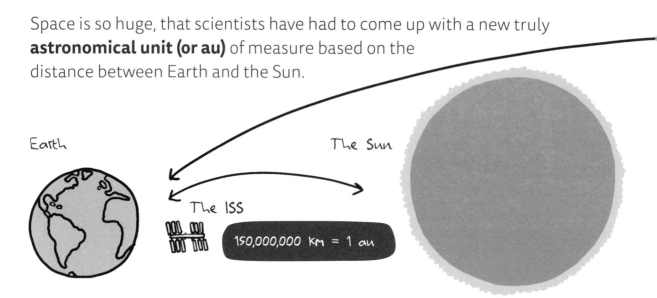

Earth

The Sun

The ISS

150,000,000 km = 1 au

And still this isn't enough to measure space.

So astronomers use light, the fastest thing we know. In space, it can travel at a speed of nearly **300,000 km per second!!!**

But to really go the distance we don't measure how long it would take to travel in seconds, we measure it in **YEARS**.

There are **31,536,000 seconds** in a year.

That means that a light year is...

## 9,461,000,000,000 kilometres long!

Let that sink in for a moment.

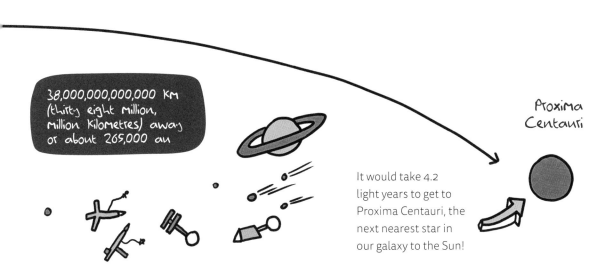

38,000,000,000,000 km
(thirty eight million,
million kilometres) away
or about 265,000 au

Proxima
Centauri

It would take 4.2
light years to get to
Proxima Centauri, the
next nearest star in
our galaxy to the Sun!

# Rock your body

But distance is not the only challenge in space travel, far from it... there are many things that can affect our bodies along the way.

Think of our **atmosphere** as a nice snuggly blanket that keeps us safe. Away from it, space is not a nice place to be.

**Space radiation** is very dangerous and would make us very sick.

Whatever way we travel in space, we are going to need a lot of protection, like a space suit or space ship that can stop the radiation getting through.

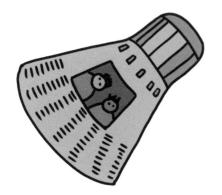

The downside is that in order to be safe in space, you pretty much have to **stay in 'lockdown'** the whole time.

Your spaceship **becomes your entire world**, and it needs to be constantly monitored in order to stay safe and to stop harmful microbes from spreading.

Lack of gravity is also not something to laugh about...

Our whole bodies are not used to **being weightless**, and being in space for a long time can weaken muscles and bones. The fluids in our bodies also shift to our heads and can even change the shape of our eyeballs!

And the longer we're away, the more important certain things become.

For example, **recycling** is not just for earthlings, it's also absolutely essential when travelling for a long time in an enclosed space.

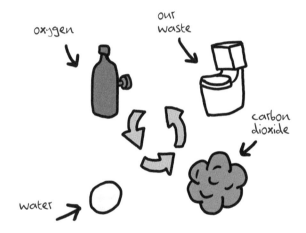

oxygen

our waste

carbon dioxide

water

## Planning, planning, planning

A trip to Mars might take three years – imagine having to think of all the situations you might encounter during that time, knowing that you can't just go back to Earth at a moment's notice!

# Have space, will travel

Imagine you are getting ready
for your big trip into space...

Draw your spaceship and your shipmates

What would your
spaceship look like?

How many rooms
will it have?

Who would come with you
on your space adventure?

Which of your favourite things would you like to take with you?

What would you miss most from Earth?

# 3, 2, 1, lift off!!

This is ground control to major you! Are you ready to go on your very own space odyssey? Time to get your imagination helmet and get inventing!

My space travel inspiration

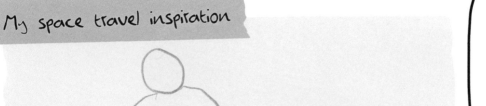

Imagine that your bedroom is a spaceship, how could you fit in everything you need?

MY INVENTION

Name it

How it works

**MY INVENTION**

Draw BIG, use colours and add labels!

Now share it on littleinventors.org!

# Wow, look who's flying!

You are ready to go to new heights!

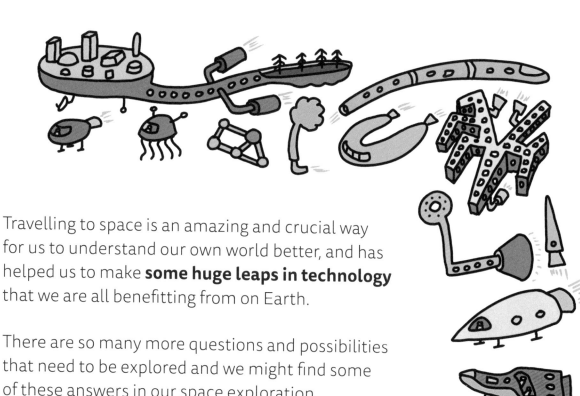

Travelling to space is an amazing and crucial way for us to understand our own world better, and has helped us to make **some huge leaps in technology** that we are all benefitting from on Earth.

There are so many more questions and possibilities that need to be explored and we might find some of these answers in our space exploration.

We might still be a very long way from hopping to the Moon for the weekend, and **travelling into space remains the privilege of the very few**.

But this doesn't mean that it will not become a reality within your lifetime!

So what about you? Would you hop on a rocket to Mars if you could?

# Fuzzy PJ fun dress up spacesuit

**Rachel, age 11**
Calgary, Canada

*"It is fuzzy on the inside. It's a real spacesuit. You can take off the helmet (in the rocket). Lightweight gloves have fake claws. Basically space PJs you can exit the ship in."*

Imagine doing a moonwalk from the comfort of your own pyjamas!

## ...to the skills of our Magnificent Makers

**Robin Ritter**, a glass blower from Ontario, Canada hand crafted this out of glass, and lined it with cosy fluffy fabric. She even knitted a miniature version of Canadian astronaut David Saint-Jacques to go inside!

Uncanny!

Photo by Canadian Space Agency

Photos by Robin Ritter

# Survival designer!

Very simply, without a spacesuit, we wouldn't be able to **survive space travel**.

Because of all the technology involved in them, spacesuits can be incredibly heavy, and that can really stop astronauts from moving freely, even in microgravity.

Each suit is designed for a specific person and mission. Parts might be **3D printed or hand sewn**. To bring their spacesuit concept to life, a spacesuit designer will work closely with a team of people with lots of different skills to produce these unique all-in-one life-supporting systems.

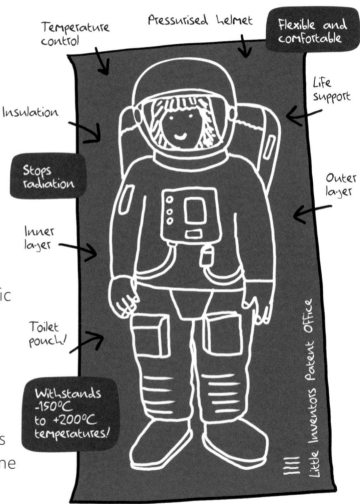

Temperature control

Pressurised helmet

Flexible and comfortable

Insulation

Life support

Stops radiation

Inner layer

Outer layer

Toilet pouch!

Withstands -150°C to +200°C temperatures!

Little Inventors Patent Office

## Chapter 7

Sleep, eat, work, play...

# Living in orbit

# The view from above...

We might dream of going into space one day... but then some people actually do it!

**Astronauts** go into space to explore it and to push the limits of our understanding of the universe.

They are able to do things that would be **impossible on solid ground** in really advanced subjects like biology, technology, space science, and physics.

They also get the most amazing view of planet Earth. It doesn't get much more exciting than that.

**Over 18,000 people apply to NASA** every year to become astronauts... but only a few are ever successful, with fewer than 600 people making it into space since Yuri Gagarin in 1961.

Want to be an astronaut? You will need to:

Speak English and Russian as a starter

Work with scientists to do research missions

HELLO! ПРИВЕТ!

Be super healthy and physically fit

Be ready to live far away from home for months, maybe more!

Train for at least two years to become an astronaut, and two more before you can go into space!

Be good at fixing things and solving problems

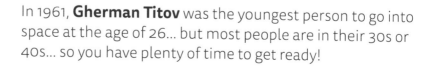

In 1961, **Gherman Titov** was the youngest person to go into space at the age of 26… but most people are in their 30s or 40s… so you have plenty of time to get ready!

# Micro, not zero

Living in orbit can be very different to being at home. For a start, **microgravity** makes people and objects look like they float in space.

**Gravity** is an invisible force that pulls two objects together. The bigger the object, the bigger the force.

Moon

Earth

People often believe there is **zero gravity** in space, but there are actually small levels of gravity everywhere in the universe.

These **keep the planets in orbit around the Sun**. The planets are actually free falling around the Sun at enormous speed – that's what being in orbit means!

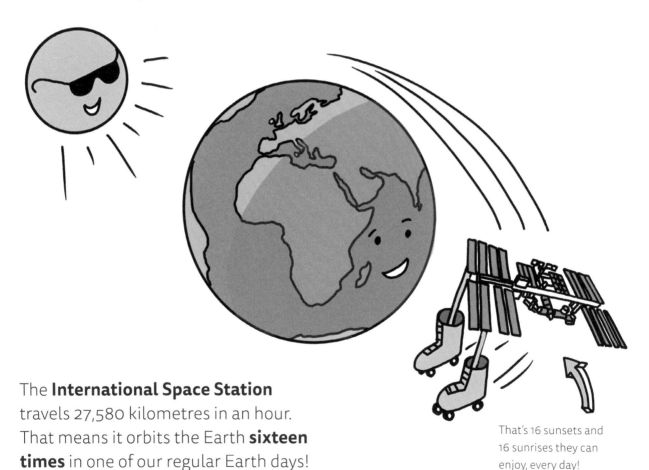

The **International Space Station** travels 27,580 kilometres in an hour. That means it orbits the Earth **sixteen times** in one of our regular Earth days!

That's 16 sunsets and 16 sunrises they can enjoy, every day!

# What is it like living in space?

Each 'day' on the International Space Station is **90 minutes long**! But the astronauts keep a normal Earth routine to their day, following Greenwich Mean Time. They even have weekends like we do!

Time for a nap! But how do you sleep in space when there is no up or down?

Astronauts have to tie themselves into a cabin to stay put. There's no lying down!

**Eating well** is vital for astronauts to stay healthy... but crumbs or liquids cannot be allowed to escape or they would risk getting into all the nooks and crannies and cause damage to the station!

And can you imagine having cereals for breakfast in the International Space Station?

**Sending anything into space**
is expensive, so everything must
be compact, long lasting,
and if it's food, tasty too!

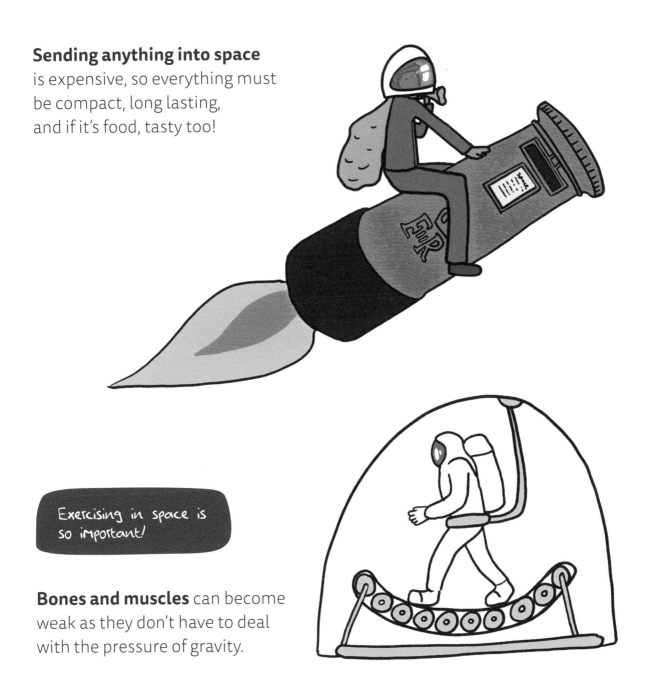

Exercising in space is
so important!

**Bones and muscles** can become
weak as they don't have to deal
with the pressure of gravity.

# Bounce! Bounce! And relax...

Once sleeping, eating, working , cleaning, maintaining the station and exercising is done, it still leaves plenty of time for fun!

How would you fill your time in the space station?

Everything is weightless – can you imagine what games you could play?

Write your Ideas

# Time to live it up!

Now that you know what it's like living in microgravity, it's time to take things to the next level and invent something that could help astronauts in space every day.

Your body might be weightless in space, but you can let your mind wander right here on Earth.

What will your brain dream up?

My space living inspiration

 Name it

_____

How it works

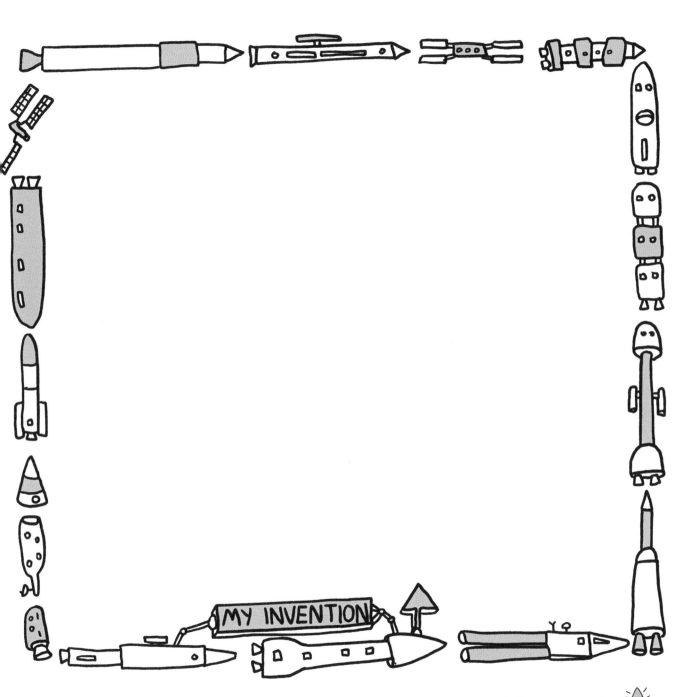

MY INVENTION

Draw BIG, use colours and add labels!

Now share it on littleinventors.org!

# Sa-*turn* that frown upside down!

SPACE SUPER STAR

Living in space is quite literally turning your everyday upside down!

It has its challenges, but each one is just **another perfect excuse to get inventing!**

This is the most extraordinary journey that a human can take – and who knows, maybe in the not-so-distant future, as space tourism develops, many more of us will get the chance to visit planets near or far, so we might as well get ready now!

# From the minds of our Little Inventors...

# The gaspasser 3000

**Emily, age 12**
Kingston, Canada

*"Ever hate having to make the area you are in stink because of having to pass gas? Well, the gaspasser 3000 is handy for that. All you have to do is when you feel ready, take the gaspasser and put the green sucker near your rear end and fire. Oh, don't forget to press the yellow button. This will turn your gas into rose-smelling scent!"*

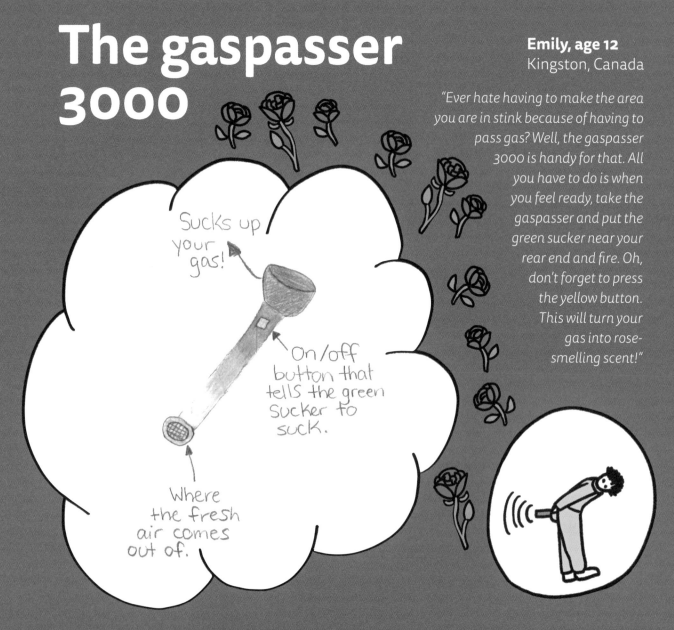

This was made real by **Diane Minier**, a designer for the **Canadian Space Agency!**

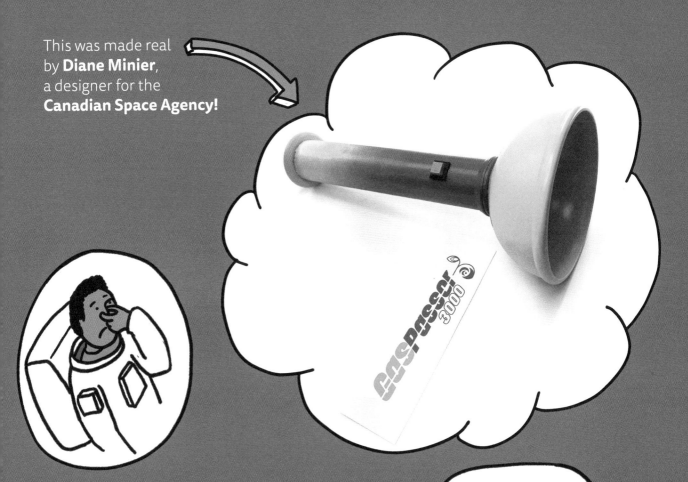

**Mathieu Caron**, Senior Operations Manager at the Canadian Space Agency thought it was a clever idea for people who work in a confined space with a team!

"It's also smart that your invention doesn't use water as water supplies in space are very limited."

# Chief sniffer!

One person who makes sure life in orbit is safe is... a 'nasalnaut'! Their job is to sniff any object that is going to be part of the International Space Station.

The ISS is a **very small, enclosed environment**, so an unpleasant or harmful smell can actually cause headaches, nausea, or can even make astronauts feel sleepy!

Plus an odd smell can be one of the first signs that **something serious is wrong** on board, so it's not a job to turn your nose up at!

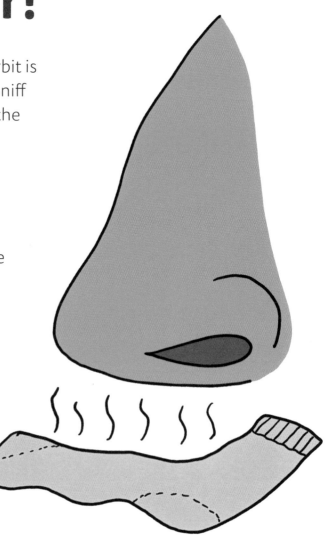

## Chapter 8

Bouncing fun, lunar lab, another step for mankind...

# Destination Moon

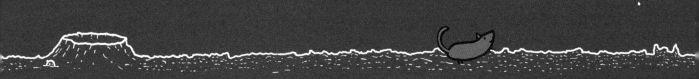

# Good Moon rising!

As the brightest and largest object in our night sky, the Moon is maybe our most familiar link to space.

It's part of our everyday lives: we can see it in all its different phases, and it causes the movement of the oceans, or tides.

It's also still, to this day, **the only place in space** that humans have actually landed in over 50 years of space exploration.

And it is **a very important part of our future in space**. In fact, the next human space adventure is definitely Moon-bound – as we aim to uncover its secrets and unique resources...

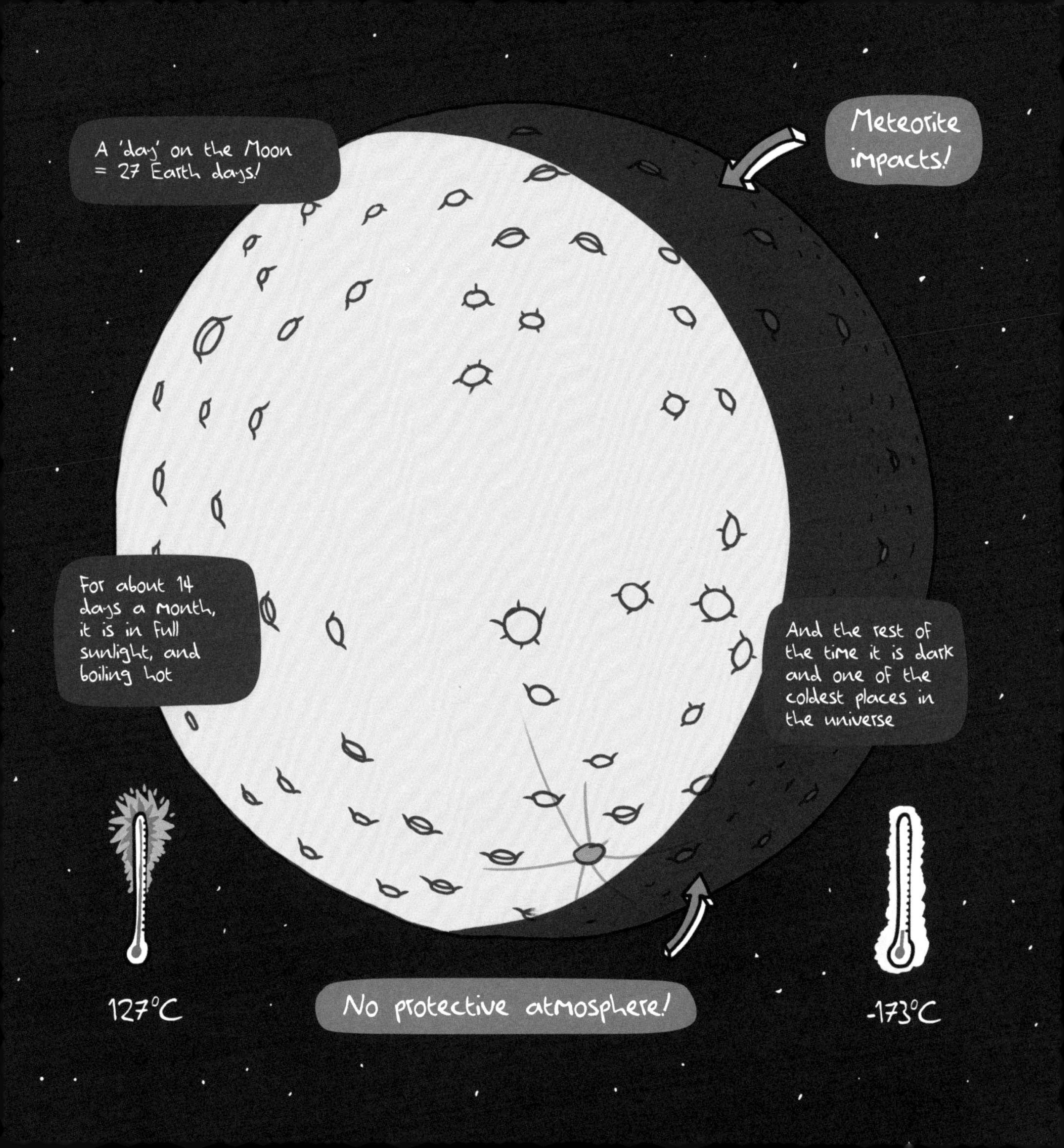

A 'day' on the Moon = 27 Earth days!
Meteorite impacts!
For about 14 days a month, it is in full sunlight, and boiling hot
And the rest of the time it is dark and one of the coldest places in the universe
127°C
No protective atmosphere!
-173°C

# Welcome to the new frontier

It has been a while since we have stepped on the Moon – but that's all about to change.

Space agencies across the world have their eyes firmly fixed on setting up a **human base** and landing the first woman on the Moon within a matter of years!

The International Space Station is due to stop operating by then, so a **lunar lab** would become the place for us humans to try, very simply, to learn to live and work away from Earth for long periods of time.

And it's not just space agencies and governments that want to go to the Moon – **private companies** are already starting to think of space as the next adventure to try on holiday.

STAR MOTEL 133

Space Hotel

**Moon dust (or regolith)** is an issue though. It covers the whole of the Moon. It is rough and electrostatic (like when you rub a balloon on your hair) so it clings to everything and can cause a lot of damage!

# Interplanetary launchpad!

Because of gravity, rockets need an enormous amount of fuel to gather enough speed to reach orbit – and the further we want to go in space, the more fuel and materials we will need.

**Setting up a base on the Moon** would then make it much easier to plan interplanetary space missions because it is in what is called **low Earth orbit** (LEO) and doesn't have as much gravity to fight against. Very simply, rockets wouldn't need as much energy to launch!

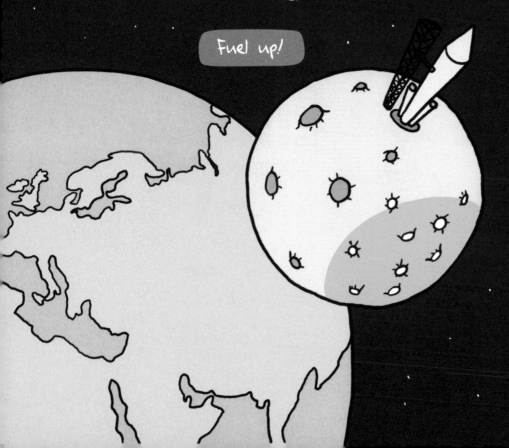

Fuel up!

Scientists have discovered that **water**, in the form of ice, exists in the shadowed craters on the south pole of the Moon!

Engineers are working on creating **robot space shuttles** that could transport items to the Moon at a fraction of the cost of human rocket missions.

The Moon itself is made of **many of the same elements** as Earth, and it could be mined for fuel and for metals or other materials.

Iron

Silicon

Rare Earth metals used for modern electronics like smartphones

One big challenge with the **lack of atmosphere** is how to protect a lunar base from a possible meteorite impact.

Scientists are still hoping to discover **how the Moon and the Earth came to be,** and more Moon exploration could help provide some answers!

# Moonday, Tuesday, Wednesday...

**Imagine it's 2030** and you are one of the first humans to live and work on the Moon... imagine what a normal day might be like!

What job would you want to do there?

Write or draw your ideas

What form of transport might you use to get around?

Imagine the kind of places you would go to for fun...

What would be more difficult to do, or easier?

# Dare you to the Moon and back!

The Moon is one of the hardest places for humans to live and we are going to need some topnotch inventing to make it work... on your marks, get set, think!

My Moon inspiration

What's a problem? It's just an invention waiting to happen!

Name it

How it works

MY INVENTION

Draw BIG, use colours and add labels!

Now share it on littleinventors.org!

# You have definitely landed!

You must truly be over the Moon!
Or under?
Who knows?!

Once in a blue moon, going on honeymoon, over the moon...
there are so many everyday expressions involving the Moon!

These show how much it has **captured the imagination** for
centuries – in songs, stories and films too!

It's really easy to look up at this familiar 'face' in the sky and start
dreaming. But it is very likely that within just a few years, going to the
Moon will be **part of our human reality**.

Though the big question is, can we learn from our experience on
Earth and respect the Moon's environment and **preserve its beauty
and magic?**

Can we become **responsible space dwellers**
before taking on the rest of the universe?

That is our next big challenge...

# The Moon coaster

**Brooke, age 12**
Vaudreuil-Dorion, Canada

"It is a roller coaster that goes around the Moon. Basically it could be used for entertainment or just to get from one place to another. I don't know if it could work but to make it you can use magnets."

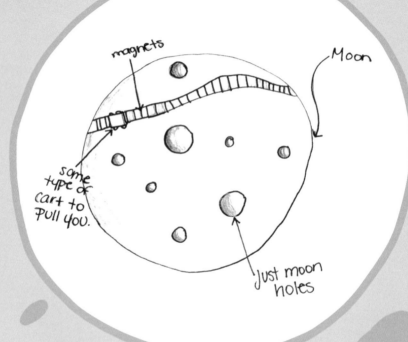

## ...to the skills of our Magnificent Makers

Our very own **Chief Inventor Dominic** created a model of Brooke's invention!

*"I thought that Brooke's Moon coaster was a great way to get around the Moon while also having lots of fun."*

*"I wanted to make the Moon rotate so I made a mechanism with little pulleys, brass rods and a handle to activate the movement.*

*I used a polystyrene ball and plaster to create a Moon texture. I cut out tiny paper circles covered in plaster to create craters."*

See it in action on littleinventors.org.

# Moon flyer!

Imagine having to navigate a hugely heavy rocket safely through the atmosphere but also in low Earth orbit – and being responsible for the safety of your fellow Moon dwellers... that's a lot of pressure!

Space pilots need to have a LOT of experience of flying on Earth before taking off **into the stratosphere**.

A normal aeroplane pilot might train in a flight simulator twice a year, but for space travel, it's **twice a DAY!**

After all, you want to be ready for any situation out there...

## Chapter 9

From a small step to a giant leap...

# Is there life on Mars?

I'm floating in a most peculiar way

# A new home?

Like all planets in our galaxy, Earth is subject to change.

Scientists believe **that one day, life on Earth will no longer be possible**. There is no need to panic, as it is unlikely to happen for hundreds of millions of years or more, but it's something that is worth thinking about.

Planning to **create human colonies on other planets** is a way to ensure we can survive as a species in the (very) long run.

Once we have our human presence on the Moon and we are ready to continue our space exploration, the next most obvious destination is **Mars**.

But we need to learn a whole lot more about it before we are ready to take our first steps on the red planet.

Some scientists believe it was an **asteroid crash** that caused the extinction of dinosaurs, and it could happen again...

In **4-5 billion years**, the Sun will get much bigger as it reaches the end of its life and could swallow the Earth...

Hundreds of millions of years from now...

...or a **mega volcanic eruption** could threaten our planet!

...and even if it didn't, we wouldn't have our Sun any more to support life on Earth.

**Climate change** could also make life on Earth more and more difficult as the planet heats up.

# The red planet

It makes a lot of sense for us to try and get to Mars.

About half the size of Earth, it's a 'close' neighbour and the planet most like ours in the Solar System. We have been exploring it with the use of probes both in orbit and on the ground for many years now.

What space probes have discovered is that long ago Mars definitely had water, a warmer, thicker atmosphere and could once have been a very good place for life to exist. But over billions of years, much of its atmosphere escaped into space, changing Mars into the dry and cold planet we know of today.

Very fine dust means the air also looks orange

Rust (iron oxide) in the soil and rocks explains its red colour

Gravity is less strong, so you could make giant leaps!

It's very cold and dry there, about -60°C on average, the same temperature as the coldest winter in Antarctica!

In fact, scientists are still searching for **any evidence of life on Mars**, either now or from the past. They also want to make a better 'map' of the surface and piece together how it evolved.

3 times taller than Mount Everest!

Olympus Mons is the highest mountain in the galaxy, at 24 km high!

A day is about the same length as on Earth, but a year is almost twice as long. That's a long time to wait for your birthday!

There are often huge, very destructive dust storms

There is not enough carbon dioxide for plants to grow!

This dust could maybe be used as a material!!

But it is likely that we won't begin to get answers until they can actually land on Mars and explore it themselves.

# Under pressure

Just like on the Moon, we wouldn't be able to just go to Mars and start living there. Because Mars is several months' travel away, we would need to find a way to sustain ourselves without having to get supplies sent from Earth all the time.

We would need to live in pressure suits and houses

The **atmospheric pressure** is too low for humans to survive in – our saliva and blood would instantly start to boil!

We would need to find clever ways to recycle water

There is only **limited water** available, in the form of solid ice.

Ice

Water

We would need to find ways to grow food

The soil is so **toxic** that plants and animals wouldn't survive.

So to create a planet fit for humans to live on, we would have to find ways to transform Mars quite radically – this is called **terraforming**.

And while we have many **greenhouse gasses and carbon dioxide** here on Earth, they would be essential on Mars to encourage an atmosphere that is similar to ours.

In the same way that they trap heat in our atmosphere, they could help to warm Mars up and protect us from the Sun's radiation.

But at present there is not enough carbon dioxide available to make this happen and unfortunately we can't take ours there!

# Plan your planet!

But imagine that scientists have managed to solve the basic problems of maintaining life on Mars... Can you create your very own Martian city?

Draw your own city

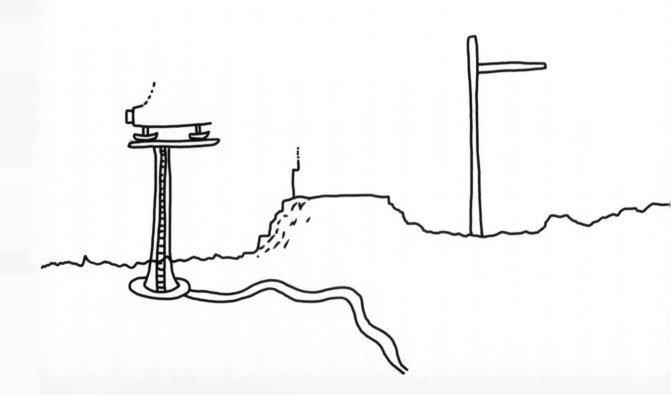

What would your home look like?

Where would you get food?

Where would you meet your friends?

## Make it *Mars-vellous!*

When there is so much stacked against something happening, the best thing to do is to let your imagination run properly wild! So what do you think could help with life on Mars?

chief Inventor

To let ideas grow, you need to give them space to breathe and feed them lots of interesting facts!

My Mars inspiration

Name it

How it works

MY INVENTION

Draw BIG, use colours and add labels!

Now share it on littleinventors.org!

# Red carpet time for you!

You've taken the red planet by storm and earned your inventing stars...

MAGNIFICENT MARTIAN

We don't know if there is any life on Mars or if there could be life there in the future…

….But there is a theory that we are all originally from Mars.

Some scientists are suggesting that a meteorite from Mars could have landed on Earth and brought with it the very first elements that made life possible.

If that is true, then **we are all Martians!**

Landing on Earth would have been a stroke of luck since this planet is much more welcoming now than Mars is!

The only way we will find out for sure is to go there…

So watch this space!

# From the minds of our Little Inventors...

# U.V. plantation

**Joshua, age 11**
Nepean, Canada

We've seen many space plants growing in films, but this is one of the crucial challenges to allow us to live in space, so we wouldn't have to take everything we needed with us!

*"It's an invention with a plant inside to give the astronauts more oxygen. It uses the U.V. rays of the sun and the water tanks attached to it and a conveyor belt."*

## ...to the skills of our Magnificent Makers

Spinning into life thanks to **Aquila Magazine** Editor and Head Maker, **Freya Hardy**!

*"This making was both very fun and very hard!*

*I started by making a ring-shaped track in a hat box. The plant capsules are attached to castors so they travel around the track, a bit like horses on a carousel.*

*I added a battery-powered motor that is normally used for spinning garden ornaments!*

*I then covered it with a massive plastic cloche (the type you can use to cover seedlings).*

*I wanted to make it look like a kind of space plantation, with little astronauts moving around."*

# Life form explorer!

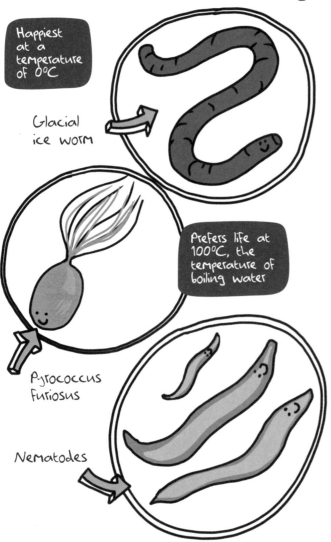

Happiest at a temperature of 0°C

Glacial ice worm

Prefers life at 100°C, the temperature of boiling water

Pyrococcus furiosus

Nematodes

One way to understand how life started on Earth or beyond, is to look at extreme places and see how life survives there, even if it can't be seen by the naked eye!

It might mean finding out everything about the **Glacial ice worm**, or microbe **Pyrococcus furiosus** (which means rushing fireball!), showing that life can indeed thrive in unlikely places, and telling us that the search for life in space is definitely worth pursuing.

Scientists are even sending **Nematodes**, a kind of worm, into space, to see how well life can adapt!

## Chapter 10

Fact finding, star exploring, future thinking...

# I'm a space Little Inventor!

**Congratulations**, you have now completed the Little Inventors in Space book – **what a star!**

Your ideas are truly **out of this world!**

# Share your ideas with the world!

We've shown you how to come up with ideas to explore space, now it's time to share your ideas with us – and the world!

Every time you see this icon within this book 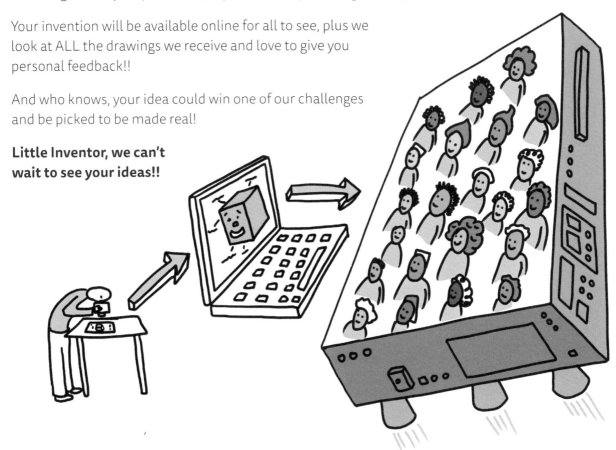 take a picture of your invention and upload it to our website **littleinventors.org** following the easy steps online (maybe with help from a grown-up).

Your invention will be available online for all to see, plus we look at ALL the drawings we receive and love to give you personal feedback!!

And who knows, your idea could win one of our challenges and be picked to be made real!

**Little Inventor, we can't wait to see your ideas!!**

# Boost your ideas up!

You had an awesome invention idea, you've drawn it and shared it with us. Is that it? Of course not!

Just like space, your imagination and ideas are limitless, they can always be improved or lead to new ideas!

## Be an internaut

Go on the Internet to see if anyone has done anything a bit like your idea. How is yours different? Or how can you change it to make it unique? Inventing is often **about improving something that already exists**, so what's your take on it?

## Sleep on it

Not literally! But take the time to forget about your idea and come back to it with a **fresh pair of eyes**. What would you do differently (or not!)?

www.littleinventors.org

## Get making

Professional makers and designers, like you, start working on their ideas by drawing them – but before making it real, they **create a model** to test it!

You can do the same by using recycled materials from around the house.

## Future forward

Imagine that someone (or something!) has been using your invention for a while. **What new use could they find for it?** Or what if someone completely different used it?

# Once upon an invention...

Having an invention idea is great, but now can you think what happens next? This is your chance to create your own invention comic!

Give your comic a name!

Who's in your story?

Where and when does it take place?

What happens with your invention?

THE END!

Or is it? Is there more to your story? Maybe it's the start of a much bigger adventure! You could also write it instead!!

# Inventor's log

And then what about all the other ideas bubbling around in your brain? You can keep a list here so you don't forget about them!

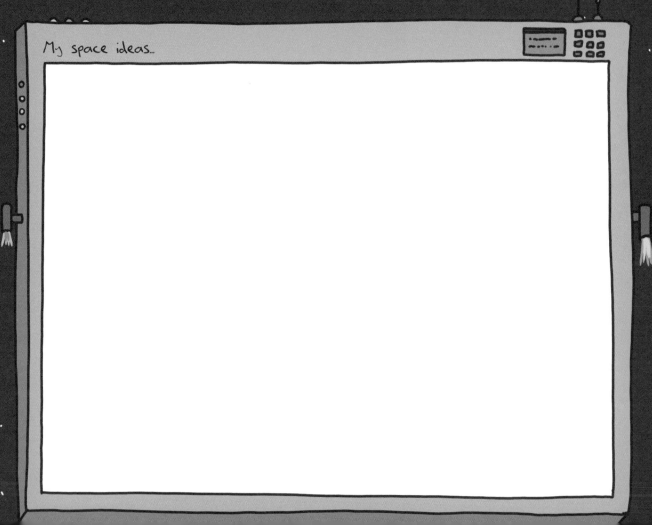

My space ideas...

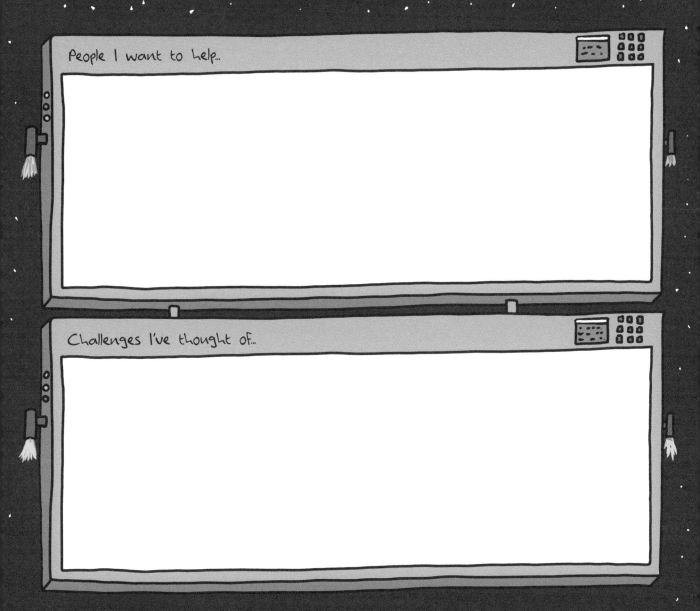

People I want to help...

Challenges I've thought of...

For all your new ideas, you can download more invention sheets from
**littleinventors.org**. We add new challenges all the time too!

# More, more, I implore!

What's that? You want more? Have you read our other books yet?

## The Little Inventors Handbook

A **step-by-step guide** to thinking up fantastical, funny or perfectly practical inventions with no limits!

## Little Inventors Go Green!

Get inventing for a **greener planet!** This activity book is full of ideas to help you come up with new inventions to make our world better.

You can also **download inventors' logs** to help you develop your idea and to make your idea into a model, how cool is that?

And get yourself over to **littleinventors.org** – you'll find mini-challenges, big challenges, and you can even submit your own!!

## Chapter 11

Why inventing is great...

# Even if you're a grown-up

# The stars of tomorrow...

Space is a big topic, and it can capture people's imaginations in so many ways, **from the scientific-minded to the daydreamers**.

The beauty is that there is room for all, and no strict rules to follow.

The imaginations of children, like space itself, are limitless, and the more they are **encouraged to explore and push the boundaries of their ideas**, the more they will learn to approach the world with an open and curious mind.

By **nurturing children's creativity and confidence** in their own ideas, you can give them the key to a brighter, wider, funnier future.

As they grow up, the children of today will most likely live in a world where space plays a much more prominent role than during our own lifetime.

Working and maybe living in space **will become a reality for many**. And who knows what discoveries await us as we get to explore further and further into our galaxy and beyond – the way we see the world could be changed entirely!

It is deeply human to have dreams bigger than ourselves and to constantly push the limit of what is 'normal'. Yet as adults, we tend to be more **afraid** to try things, to make mistakes, to be wrong.

But for children, everything is new, fresh, exciting. Just for a moment, **suspend your sense of what is or isn't possible**.

Remember that imagination is all that it takes to start to make things happen – so enjoy the fantastical, thoughtful or plain bonkers ideas dreamt up by your Little Inventors.

After all, they are the holders of tomorrow, and you can be part of **letting them shine as brightly then as they do now**.

# Top tips for grown-ups

## Suspend belief...

Invention is about exploring ideas freely, so don't worry if the laws of physics are being challenged!

## It's all good!

Ideas can be big or small, but to your Little Inventor, they will be new and exciting! Share their enthusiasm and keep them going!

## Feed their curiosity

Learning doesn't have to be boring. Find the little facts that create sparks and awaken their appetite for more!

## Who's in charge?

When it comes to imagination, it's a good idea to let your child lead the way...

## Look up!

The night sky is right there, for all to see – take the time to wonder at everything happening there – stars forming, satellites whirring, other worlds living...

## Put yourself in their place

We take so many things for granted, but children don't – rediscover the world through their eyes!